BUILDING
WORLD LANDMARKS

The
Chunnel

by Joanne Mattern

BLACKBIRCH®
PRESS

THOMSON

GALE

San Diego • Detroit • New York • San Francisco • Cleveland • New Haven, Conn. • Waterville, Maine • London • Munich

THOMSON

✳

GALE

For more information, contact
The Gale Group, Inc.
27500 Drake Rd.
Farmington Hills, MI 48331-3535
Or you can visit our Internet site at http://www.gale.com

LIBRARY OF CONGRESS CATALOGING-IN-PUBLICATION DATA

Mattern, Joanne, 1963-
  The Chunnel / by Joanne Mattern.
    p. cm. — (Building world landmarks)
  ISBN 1-56711-301-X (hardback : alk. paper)
  1. Channel Tunnel (England and France) I. Title. II. Series.

  TF238.C4M3 2004
  624.1'94'0916336--dc21

                                                                2003008047

Table of Contents

Crossing the Gap

THE BRITISH ISLES are separated from the rest of Europe by the North Sea to the east and the English Channel to the south. The English Channel flows between England and France and has protected England from invasion by foreign armies for more than one thousand years. This separation gave the tiny island nation a strong sense of identity. It also, however, made it difficult to travel or ship goods from England to the rest of Europe.

For hundreds of years, people tried to think of a way to link England and France across the channel. Since the channel is only twenty-two miles wide at its narrowest point, creating a link did not seem too difficult in theory. Some people wanted to build a bridge, while others dreamed of a tunnel drilled through the rock under the channel bed. Some people came up with plans for a combination of both. Although plans

Opposite:
The English Channel, shown in this seventeenth-century map, separates England and France by only twenty-two miles at its narrowest point.

5

Over the years, many people came up with different ways to link England and France. This 1857 design proposes an underwater tunnel with ventilation shafts that rise above the water's surface.

were drawn up several times and several tunnels were actually begun, none of these projects was ever completed. Sometimes, the projects were stopped because they were too expensive to build. Other times, the strong desire of the English people to remain separated from Europe forced the plans to be abandoned.

Finally, in 1984, England and France agreed to build a tunnel. Construction was started in late 1987 and continued for six years. As with many construction projects, many complications and challenges arose during the project. Engineers and workers had to worry about the thickness of the rock under the water and how to carry equipment in and waste material out. They had to make sure the tunnel was safe and able to handle the many trains that would travel through it each day. They had to figure out ways to keep the tunnel from getting too hot or too wet. Since two crews worked at the same time—one started in England and

the other started in France—there was great fear that the two projects might not meet in the middle.

In 1993, the chunnel was completed. A new age of travel had arrived in Europe. The tunnel crossing took only half an hour and was free of choppy seas and bad weather. Queen Elizabeth II of England called the chunnel "the engineering marvel of the century." Its construction changed the map of Europe and the lives of the people who live there.

When the chunnel opened in 1994, travel between England and the rest of Europe became fast, comfortable, and easy.

7

Chapter 1

The Tunnel Nobody Wanted

PEOPLE HAVE BEEN trying to find an easy way to cross the English Channel for hundreds of years. The channel is not very wide, but it has proved to be quite a barrier to travel between England and France. For centuries, ships were the only way to cross the channel. This method, however, was difficult due to the extremely choppy water in the channel and frequent rainy and windy weather. Thick fogs often settle over the water and make visibility poor.

Another problem with the channel is what lies at its bottom. Many tall ridges of sand rise from the channel's floor. These ridges create fast, tricky currents and make some parts of the channel very dangerous. There are not many safe harbors on either side of the channel either. These conditions combine with bad weather to make the crossing a miserable experience.

Opposite:
Construction of a tunnel began in the 1970s, but after only a half mile was dug on the English side, financial problems stopped the project.

In addition to hampering travel, the channel also created a barrier to trade. Goods could be shipped through the rest of Europe by train or truck, but all goods that were sent out of or into England had to be delivered by ship or airplane. A tunnel or bridge that connected England to France would provide a quick, efficient, and financially sensible method of transporting goods to and from the island nation.

Arguments Against a Tunnel

Despite the potential convenience of having a tunnel link to Europe, English people were vehemently opposed to the idea for centuries. The most important reason was a fear of invasion. For thousands of years, the English Channel has protected Great Britain the way a moat protects a castle. Only two armies have ever been able to cross the English Channel and invade Great Britain. The first was the Roman army in 54 B.C. The second invasion occurred in 1066, when William of Normandy and his army crossed into England from France.

During World War II (1939–1945), most of Europe came under the control of Adolf Hitler's German army. In 1940, Germany invaded France and occupied the country until 1944. It was no secret that Hitler wanted to invade England. If a tunnel had existed across the channel at that time, it would have been easy for the German army to march from occupied France into England. There was no tunnel, though, and Hitler was never able to conquer Great Britain. For years afterward, people who opposed a tunnel supported their

Although Hitler occupied France for four years during World War II, he was unable to invade England because there was no easy way to cross the English Channel.

viewpoint with the fact that the English Channel had protected England during World War II.

The lack of a channel crossing also protected England from disease. Millions of animals in Europe have died from rabies. If a person is bitten by a rabid animal, he or she will also catch the disease, which is almost always fatal. England, however, has not had a case of rabies in more than one hundred years. Because it is an island, it is almost impossible for wild animals to come into the country. Domestic animals, such as dogs and cats, must have proof of a rabies vaccination or be quarantined to be sure they are free of the disease before they are allowed into England.

A sense of national pride was another important reason the British people resisted the idea of a tunnel across the English Channel. Isolation from the rest of Europe has allowed Great Britain to develop a distinct character. Many British people do not like or trust other European countries. This distrust and rivalry is especially strong between England and France.

Early Efforts to Build a Tunnel

The English people may not have wanted a tunnel, but the French certainly did. One of the first proposals for a tunnel was created in 1802. At that time, Napoléon Bonaparte, the leader of France, asked a

An attempt to build a tunnel in the 1880s sparked lively debate. After workers had dug about 11,500 feet, the British people convinced the government to halt construction.

French engineer named Albert Mathieu to come up with a plan.

Mathieu drew a plan for a tunnel with two tubes. One tube would carry horsedrawn carriages, while the other, smaller tube would provide ventilation. Although Napoléon was enthusiastic about the project, the British government refused. The fear of an invasion was just too strong. Mathieu's tunnel was never built.

During the mid-1800s, a Frenchman named Aime Thome de Gamond came up with many plans for a channel crossing. Gamond's plans included bridges, tunnels, and combinations of the two. In 1856, he drew up a plan that called for an artificial island to be built in the middle of the channel. This island would provide a port for ships, as well as a train station. Both the station and the port would connect to the tunnel by a spiral ramp. Most of Gamond's plans were not practical, and none was ever used to build a crossing.

In the 1880s, both the French and British governments agreed on a tunnel plan. For the first time, men began to dig off the coasts of England and France. A tunnel boring machine had recently been invented, and the machine made the construction easier. Crews were able to dig out about 328 feet per week. Plans called for the tunnel to be completed in about six years.

This project, however, was doomed. Once again, the British people worried that a European army would have an easy time invading England through the tunnel. Many people signed petitions and called for the government to stop the project. Ferry companies also complained, because they were afraid they would lose all their business if people could cross the channel by tunnel instead.

Newspapers and magazines printed articles for and against the project, creating a lively debate that went on for months. Supporters of the tunnel said that the project would put an end to miserable sea crossings and improve business between England and the rest of Europe. Eventually, public opinion won out. The British government listened to the group who opposed the tunnel and changed its mind about the project. In 1882, Great Britain refused to give final approval to the project. About 11,500 feet had been dug out, but the project could not go forward without Great Britain's approval.

Although people continued to discuss possible channel crossings, almost one hundred years passed before another attempt was made to link England and France. In 1973, the two countries agreed on a tunnel project. Work began near Dover, England, but only half a mile was dug before Great Britain pulled out of the deal. This time the reason was not a fear of invasion, but financial problems. England was suffering from a weak economy. Prices were high, and many people were out of work. The government no longer felt it could spend money on the construction project. Then France refused to allow England to join the European Common Market, an organization that made it easier for European countries to trade with each other. That was the last straw. Great Britain pulled out of the project in 1975. Once again, it looked as if a channel crossing would never be built.

The English Channel

The English Channel is a narrow body of water that separates England from France. It is located off the southern coast of England and is about 350 miles long. At its widest point, the channel is 150 miles across, but the part of the channel called the Strait of Dover is only 22 miles wide. The Strait of Dover separates Dover, England, and Calais, France, and is the point where all attempts to cross the channel by tunnel or bridge have been made.

The English Channel links the Atlantic Ocean in the west with the

As seen in this view from space, the English Channel separates southern England from the Normandy region of France.

North Sea in the east. There are several islands in the channel, including the Channel Islands near France, and the Isle of Wight, off the coast of England. The channel was probably created about five hundred thousand years ago, when the North Sea overflowed and flooded a land bridge between England and France.

Chapter 2

What Will the Project Be?

IN 1984, THE world heard stunning news: British prime minister Margaret Thatcher and French president Francois Mitterrand announced that their two countries wanted to build a link across the channel. They said this project "would be in the mutual interests of both countries."[1]

The French government had wanted to build a crossing for more than one hundred years, but England had always opposed the idea. During the 1980s, however, Great Britain and the rest of Europe were in the middle of an economic crisis. Thatcher knew that a project to cross the channel would create thousands of jobs and thereby pump money into the British economy, especially in the towns near the construction site. Moreover, English factories would produce materials for the construction, which would also benefit the country's economy.

Opposite:
In 1984, French prime minister Francois Mitterrand (left) and British prime minister Margaret Thatcher (center) announced that their countries would build a link across the English Channel.

Despite the long-term economic advantages, Thatcher knew that the channel project would be expensive, especially in the early stages. She did not want to tie up England's money in the construction project and saddle the country with more debt. For this reason, Thatcher stated that no government money would fund the project. All the money would have to be raised by the builders themselves.

Several builders were anxious to take on the challenge. They could raise money by selling stocks, or shares in the company, to the public. People would buy the stocks as an investment, in the hope they would be worth more money once the crossing was completed. As the company made money through ticket sales or tolls, the value of its stock would go up, and the investors would make money too. The builders could also borrow money from banks to pay for the project. Although the financial investment was tremendous, the builders were confident that they could handle the challenge. Now they just had to find out exactly who would build the crossing, and what the project would be.

A Design Contest

The first step in crossing the channel was to choose a design. In April 1985, Thatcher and Mitterrand announced a design competition. Companies were asked to submit plans that described the project, how long it would take to build, and how much it would cost. The design could be a bridge, a tunnel, or a combination of the two. The deadline for submissions was October 31,

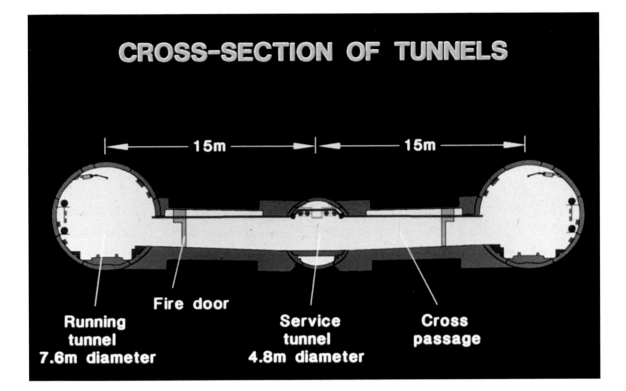

CROSS-SECTION OF TUNNELS

15m

15m

Running
tunnel
7.6m diameter

Fire door

Service
tunnel
4.8m diameter

Cross
passage

1985, and a committee would pick the winner in January 1986.

Ten proposals were submitted, but only four had enough financial support to be taken seriously. Many people liked a plan called Eurotunnel and nicknamed the "chunnel." This plan, created by a British engineering firm named Balfour Beatty Construction, was similar to the 1802 and 1973 projects and called for two railroad tunnels with a service tunnel running between them. While people could not drive through the tunnel themselves, they could transport their cars on one of the trains. This plan would cost $3.6 billion.

Another plan, called the Channel Expressway, included two tunnels large enough to carry both trains

Plans for the Eurotunnel called for two main train tunnels with a service tunnel between them.

and cars. Many people, however, thought it was unsafe to have trains and cars using the same tunnel because if a train derailed, it might crash into the cars sharing the tunnel. They also worried that the undersea rock was not thick enough for the extra-large tunnels this project called for and that the ventilation system would not work.

The British Steel Company proposed a bridge and tunnel system called the EuroRoute. This plan featured a bridge that would extend out to an artificial island in the middle of the channel. There, drivers would drive down a curved ramp into a tunnel that led to another artificial island. Then they would drive onto another bridge and finally arrive at the opposite shore. Although people liked the idea of driving their own cars across the channel, the $7 billion cost of this plan seemed too expensive.

Finally, there was a plan known as EuroBridge. This plan featured a huge bridge that held a plastic tube suspended by cables. This project was also expensive, with a cost of more than $6 billion. More importantly, many people thought it was unsafe to have such a long bridge across the channel. The large number of ships that traveled through the water meant that a ship might hit the bridge one day and cause a terrible disaster.

The Winning Idea

On January 20, 1986, the creators of all the plans and the judges all gathered in Lille, France. That afternoon, the world heard the long-awaited announcement. The winner of the design competition and the project that would link England and France across the English Channel was the Eurotunnel, or Chunnel.

Balfour Beatty Construction was thrilled at the news, but the company knew it could not create the tunnel alone. It joined with four British and five French construction companies to do the job. The company became known as the Channel Tunnel Group/France Manche. Soon, the tunnel and the company were both known simply as the Eurotunnel.

Finding a Crew

From the beginning, building the chunnel was full of challenges. Because the designers only had seven months to create their winning design, they had only included basic information. Now that they actually had to build the chunnel, they had to answer many questions.

One of the first questions was where Eurotunnel was going to find enough engineers to complete the project. Eurotunnel estimated that the project would take seven years to complete. Because most engineering projects take about two years, engineers were used to working on a project for a year or two and then traveling to another country and another job. Many

Out of ten design proposals submitted in the channel crossing competition, Balfour Beatty Construction's Eurotunnel was the winning design.

engineers were reluctant to sign onto a project that would take such a long time. To commit at least seven years to one job was almost unheard of.

The unique nature of the chunnel project also made it hard at first to convince engineers to work on it. Engineers often used the knowledge acquired from one building project to construct another project. The Eurotunnel, however, would require construction techniques that probably would never be used again.

Although engineers and workers were initially reluctant to join the Chunnel project, they soon became intrigued by the challenge of the once-in-a-lifetime opportunity.

John Hester, an American engineer who worked on the project, said, "There was nothing being done on the Channel Tunnel that's going to do any good anywhere else, so why should we pay any attention to it? You'll learn more from a little sewer job in Detroit than you will on the Channel Tunnel."[2]

Eventually, however, the tremendous challenge of building the chunnel finally won over reluctant engineers. To build a link between England and France would be a once-in-a-lifetime experience. In the end, few engineers could turn down the opportunity.

Finding workers to actually build the tunnel was not as much of a challenge as finding engineers. In England, men who had tunneling experience were hired. Although some of these men shared the engineers' reluctance to commit so many years to one job, most were thrilled at the chance to find steady work with excellent pay. Many of the British tunnelers were from Ireland. Ireland had been an important source of crews for tunnel projects since the 1940s.

The French took a different approach to hiring their crews. Instead of looking for experienced tunnelers, the French engineers wanted to hire and train workers for the job. The French government also encouraged this approach, because it provided jobs for the many unemployed men near the village of Sangatte, where the French end of the tunnel would be constructed. As John Hester described the French procedure: "That side was a factory. You take a little guy off the street, you train him, you put him in there. . . . He's taught what to do, and he does it."[3]

Inventing Technology

Once the engineers and workers were lined up, attention turned to the technical aspects of the job. It was one thing to plan to tunnel through the rock and chalk underneath the English channel. It was quite another thing to actually do it. At the time the project started, there was no machine that could handle the specific conditions required by the channel job, which required digging through a tricky mix of rock, chalk, and clay; so the engineers set about inventing one.

Eurotunnel hired an American company called the Robbins Company to build a tunnel-boring machine, or TBM. Robbins specialized in making tunneling machines, but it had never had a job this big before. It came up with a huge TBM that weighed more than fifteen thousand tons. The TBM worked like a giant drill to tunnel through the rock. The drill head was fifty feet in diameter. Robbins built eleven TBMs for the project. These TBMs were so huge that they had to be sent to the site in pieces and then put together inside the tunnel. Before they arrived, workers used conventional drills and equipment to enlarge the tunnel entrance that had been dug during the unfinished 1973 project. This area became the chamber where the TBMs were put together.

Engineers knew the TBMs that Robbins built would work fine on the English side because the rock on that side of the channel was made of chalk, which is relatively soft and easy to cut. This rock also contained a lot of clay. Because clay is waterproof, it kept the water from the channel from rushing into the tunnel. The rock on the French side, however, did not have as much clay. That meant that having water break through the tunnel walls would be a constant danger. Workers there needed a TBM that could stand the higher water pressure pushing down on the tunnel. To solve this problem, Robbins built a special TBM that was half submarine, half drill. This TBM was nicknamed Brigitte, and it would be capable of working deep underwater, just as a submarine could withstand the severe pressure of the water deep under the ocean.

Chapter 3

Building the Chunnel

WORK ON THE chunnel finally began in December 1987. The first order of business was to build the central service tunnel.

Early Progress

The tunnel crews started to dig at Shakespeare Cliff, near the city of Dover and about five miles from Folkestone, where the railway terminal and control center would be built. There was already a small tunnel opening under the cliffs from the failed 1973 project. On the French side, workers started to tunnel near the village of Sangatte, two miles from Coquelles Terminal, which would serve as the French rail station. As the tunnel rose from under the channel, it continued underground until it reached the rail stations.

One crew started in England and dug toward France. The other crew started in France and dug toward

Opposite:
In December 1987, one crew started digging in England and another crew began in France, with engineers constantly checking the direction of the tunnels from each side.

England. From the beginning, digging through the chalk and rock that made up the bed of the channel was a huge challenge. The TBMs could cut through about 750 feet of rock every day. This produced debris such as rock, dust, and mud. All this debris had to be taken to the surface so it would not be in the way of the workers or the tunnel. To remove the debris, a conveyor belt was built inside the tunnel. Rail cars ran to the surface along this belt, 24,000 tons of debris out of the tunnel every hour. The debris was dumped into the channel behind a newly constructed seawall. When it was finished, this debris area added 85 acres to England.

The tunnel rose from under the channel near the village of Sangatte on the French side.

The crews also had to find a way to keep water out of the tunnel. A leak could damage the equipment, flood the tunnel, and kill dozens of workers. To prevent the tunnel from filling with water, crews used hoses to pump out any water that seeped through the tunnel walls. Everything inside the tunnel, including electrical cables and wiring, had to be waterproofed. To keep the tunnel from caving in under the water pressure, extra-heavy waterproof concrete liners were installed along the tunnel walls. As the TBMs dug through the rock, crews installed the concrete liners. The concrete was mixed at the surface and sent down in special trains with rotating bodies to keep the concrete from hardening. After the concrete hardened, curved sheets of steel were installed inside the concrete. These steel sheets formed the inside walls of the tunnel.

Meeting in the Middle

As the British and French TBMs dug toward each other, engineers constantly checked their directions. Because they were digging through rock that was thicker in some spots than in others, the TBMs could not stay on a perfectly straight course. If the two tunnel halves were even a few feet out of alignment, however, the project would be delayed and millions more dollars would have to be spent to fix the problem. If the distance between the tunnels was too great, the whole project might have to be abandoned.

Miles of solid rock separated the two machines, so their operators could not see each other and keep the drills on course by sight. Some tunnel projects use a satellite mapping program to keep the two segments aligned. This would not work on the chunnel project, though, because the tunnel was too far underwater for a satellite to find it.

To solve these problems, engineers developed a laser guidance system for each TBM. A laser was mounted on each drill and sent a beam of light to a control panel in the tunnel. This control panel then sent a signal back to the TBM operators to tell them if the TBM needed to be turned slightly to stay on course. Still, no one would know if the guidance system worked until the two tunnels actually met.

On October 30, 1990, the two TBMs were only 330 feet apart. The British TBM stopped and waited for the Brigitte, the French TBM, to break through the rock separating the two machines.

Workers had placed a long, thin drill on the front of the huge British TBM. This drill served as a probe. The plan was for the probe to break through the rock ahead of the TBM so that the French workers at the other end of the tunnel would see it first and be able to figure out where the British machine was. Then Brigitte would drill straight into the British tunnel.

A British worker named Steve Cargo was onboard Brigitte. As Brigitte tunneled slowly toward the other TBM, Cargo talked to British engineers on the phone. Cargo strained to see in the darkness. Suddenly, he saw water flowing through a small hole in the rock in front of him. Then he saw the drill. The two tunnels were only twenty inches out of alignment—a small enough distance that it would not cause any problems.

Crews from England and France celebrated when, on December 1, 1990, the two sections of the central service tunnel met underneath the English Channel.

Brigitte continued to tunnel through the rock to break through to the other side. One British worker and one French worker had been chosen by lottery to be the first to meet inside the tunnel. Finally, on December 1, the two men met and shook hands underneath the English Channel, while the crews and engineers celebrated. The English Channel had finally been conquered.

One Down, Two to Go

The service tunnel was completed, but there was still a great deal of work to be done. Work began on the two main tunnels, one on each side of the service tunnel. At twenty five feet wide, these tunnels were much larger than the service tunnel, which was about thirteen feet wide. Workers also had to build connecting tunnels from the main tunnels to the service tunnels. These connections were called crossovers. They would allow trains to switch tunnels if they had to and maintenance workers and emergency personnel to reach all the tunnels quickly and safely.

As they had during the construction of the service tunnel, crews working on the main tunnels and the crossovers had to deal with the constant presence of water breaking through the tunnel walls and roof. One day, the roof of one of the crossovers began to crack. Within three hours, the steel reinforcing bars had buckled, and the roof had settled more than two-and-a-half inches. Engineers discovered that a large amount of water was resting on top of the clay above the roof and increasing the pressure on the tunnel. Workers solved the problem by drilling holes in the clay to release the water.

The two main tunnels were completed in the summer of 1991. Each main tunnel (pictured) is twenty-five feet wide.

As they constructed the main tunnels, the engineers always had to be ready for surprises. Sometimes the rock was thicker than they expected it to be, which made it harder to drill through. Other times, large cracks made tunneling through an area dangerous. In these cases, the engineers would have to figure out how to change the path of the tunnel so it followed the natural course of the rock and still keep it on target to meet the tunnel being dug on the other side of the channel.

On May 22, 1991, the northern tunnel was completed. This tunnel would carry passengers from England to France. Just over a month later, on June 28, the southern tunnel was finished. This tunnel would carry passengers from France to England.

As each tunnel was completed, the TBMs had to be removed. These machines could not travel backward, so engineers had to come up with a creative solution to get them out of the way. The answer was to drive one TBM into the rock deep underneath the tunnel, fill the hole with concrete, and leave the machine there forever. Then the other TBM could drive straight ahead to the other side, where it was taken apart and returned to the surface. Different TBMs worked on each tunnel.

Finishing Touches

Once the tunnels were completed, workers began to lay tracks for the trains. Overhead cables were installed to provide electricity to power the trains. Meanwhile, special Eurostar trains were being built in

After the tunnels were completed, workers laid train tracks (pictured), and installed ventilation systems, lights, and overhead cables.

factories aboveground. These high-speed trains could travel 186 miles per hour on the ground and about 100 miles per hour inside the tunnel. Slower speeds were required inside the tunnel in order to prevent too much air from building up in front of the moving train.

Workers also laid electrical cables inside the tunnels and installed lights and ventilation systems. Special tubes called piston relief ducts were also installed. As a train moved through the tunnel, it pushed air in front of it. The piston relief ducts allowed the air to escape and keep the air pressure in the tunnel the same from end to end.

Safety measures also had to be put into place. Engineers were well aware of the many disasters that could happen inside a tunnel, from a fire to an earthquake, a train wreck to a terrorist attack. To prevent problems, workers installed fireproof doors at intervals between the main tunnels and the service tunnel to prevent flames from spreading. They also installed a number of walkways from each tunnel to the service tunnel. Passengers and crew could use these walkways to escape in case of a disaster. In addition, the air pressure was kept higher in the central service tunnel to force smoke and fumes out and make it a safe haven in case of accidents.

Finally, in late 1993, the chunnel was completed. Its average depth was 130 feet under the channel. Now the real adventure would begin.

High-speed Eurostar trains were specially built to travel 186 miles per hour above ground and 100 miles per hour in the tunnel.

Chapter 4

A New Connection

BEFORE THE CHUNNEL could open to the public, it had to be tested. On December 10, 1993, a train that carried several hundred specially invited guests traveled from France to England. Soon after this train arrived in Folkestone, a British train carried its passengers to France. The trips through the tunnel lasted just twenty-two minutes each. Then a huge party was held to celebrate this historic event.

Workers spent the next six months putting finishing touches on the tunnels and the trains and making sure everything would run safely and smoothly. Finally, on May 6, 1994, the chunnel officially opened for business. It had cost more than $16 billion to build, which was much more than the original estimate. As often happens during construction, unforeseen circumstances make things more expensive, and the cost of materials and labor can be more than originally

Opposite:
Queen Elizabeth II and French president Francois Mitterrand cut the ceremonial ribbon at the Chunnel's 1994 opening. The Chunnel had taken almost seven years to build and cost more than $16 billion.

expected. The general feeling, though, was that the chunnel would be worth the money.

Queen Elizabeth II of Great Britain and French president Francois Mitterrand were there to ride the train and cut a ribbon to open the chunnel. They also paid tribute to the fifteen thousand men and women who had worked on the project, as well as the eleven workers who died during the construction.

Passengers on the Chunnel travel in comfort and can get from London to Paris in three hours.

A Popular Way to Travel

From the beginning, Eurotunnel was a success. People were thrilled to find a way to cross the channel with-

out seasickness and bad weather. A ride through the chunnel was also faster than a ferry crossing. The Eurostar trains could cross the channel in about thirty minutes, plus time for loading and unloading passengers and cars. The same trip by ferry took at least an hour and a half. Passengers could stay in their cars during the crossing or ride in a train car. The chunnel made it possible to travel from London to Paris in three hours. Soon a London-to-Brussels link was added. This trip took about three hours and fifteen minutes.

For the next two years, everything went smoothly underneath the channel. Then, an event occurred that put Eurotunnel's safety and evacuation systems to the test.

Disaster!

On November 18, 1996, a truck carried on a train in the southern tunnel caught fire as the train left Coquelles. Soon, five other trucks caught fire as well. The fire spread quickly, which sent smoke into the dining car of the train.

Workers on the train had practiced what to do in case of a fire. They told the passengers to lie down on the floor and wait for help. Soon French firefighters arrived and led everyone into the service tunnel, which was free from smoke and flames. Thirty-four passengers on the train escaped without serious injury and were soon on their way back to France.

Although the fire did not kill or seriously injure anyone, it caused a lot of damage to the tunnel. More than six hundred yards of the tunnel's liner was destroyed, which exposed the steel frame underneath. It took six months and more than $80 million to repair the damage. New concrete liners and train tracks were installed, along with new cooling pipes, overhead wires, and telephone connections. Train service continued during the repairs. The trains switched to the northern tunnel to pass by the closed area and then switched back to the southern tunnel.

People were alarmed to hear about the fire, but they were also reassured by the quick response of train workers and rescue crews. The evacuation had gone very smoothly, and no one was seriously hurt. It was obvious to most people that the chunnel was a very safe way to travel. It did not take long for riders to return to the Eurotunnel.

The Chunnel Today

By 1996, at least 60 percent of all passengers between England and France used Eurostar trains, and a total of 57 million people traveled through the chunnel between 1994 and 2000. During 2000, 3 million cars, 1 million trucks, and 7 million passengers used the chunnel.

The chunnel has also made it easier to transport freight. Trucks loaded with all sorts of cargo can now drive onboard the special Eurostar trains and be in France in half an hour. This is much faster and more economical than transporting goods by ship.

Passengers are not the only people who have benefited from the chunnel. The construction and maintenance of the tunnel, as well as the Folkestone and

Coquelles railway terminals, created thousands of jobs, which improved the local economies of both areas.

Problems

Despite its phenomenal success, the chunnel has created some problems. Perhaps the biggest issue is refugees. Over the years, many people have escaped from Iran, Afghanistan, Pakistan, Russia, and other countries and ended up in France. French laws concerning these illegal aliens are very strict, and many want to travel to England, where the laws are more lenient. Thousands of these refugees gather in Sangatte, near the tunnel entrance, and try to sneak through the chunnel into England.

Trucks loaded with freight can drive onto a Eurostar train in France and arrive in England thirty minutes later

Refugees will try almost anything to get through the tunnel. Some sneak onboard trucks as they drive onto the train. Others hide underneath the train or between the train cars. This is very dangerous, and many refugees have been killed.

The British government fines Eurotunnel for every illegal alien that sneaks through the chunnel. To prevent this, the company employs more than one hundred security guards. Security cameras, electric fences, and specially trained dogs also try to discourage refugees from crossing. Still, up to two hundred people try to get through the tunnel every night, desperate to find a better life.

The British also worry about other unauthorized immigrants—rats, mice, and other small mammals that might slip through the tunnel. These animals carry rabies and could introduce the disease to Great Britain. To prevent this from happening, steel screens cover every opening inside the tunnel. So far, this system has worked, and Great Britain is still free of rabies.

Terrorism is a constant fear as well. Public places in Great Britain have long been targets for the Irish Republican Army, or IRA. The IRA is a group of people who want Northern Ireland to be free of British rule. In the late 1990s, British police discovered an IRA plot to blow up the chunnel. Although this plot was stopped, the threat of terrorism—from the IRA and other groups around the world—is still a real danger. The British government constantly monitors suspicious activities and relies on security guards and high-tech computers to keep the chunnel safe.

Folkestone and Coquelles

Before passengers get into the chunnel, they start their trip in one of two railroad stations on land. In England, trains depart and arrive at the Folkestone Rail Control Center. In France, the station is the Coquelles Terminal.

Folkestone is the main rail control center. There is a special room inside the terminal that is filled with people called controllers. The controllers use a huge electronic screen to monitor everything that goes on in the tunnel. Computers are constantly sending information from the tunnel and the trains. In case of an

The chunnel's main rail control center is in the train station in Folkestone, England.

emergency, the controllers can take over the automatic operations and call for help if needed.

Coquelles also has a control center, but so far it has only been used to monitor trains going in and out of the station. The Coquelles control center was built as a backup system to the control center in Folkestone and can take over operations if something puts the Folkestone center out of commission.

Despite threats of fire and terrorism, the chunnel and its Eurostar trains remain triumphs of modern transportation.

A New Way to Travel

For hundreds of years, people dreamed of a quick, easy way to cross the English Channel. The chunnel made this dream a reality and transformed travel between England and the rest of Europe. What was once difficult and demanding is now easy and enjoyable. As with so many projects that use technology, the chunnel has created a new and better way of life.

Chronology

1802   Albert Mathieu draws up plans for a tunnel between France and England.

1856   Aime Thome de Gamond creates a plan for an artificial island to be built in the middle of the channel. The project is never attempted.

1880s   The French and British governments agree on a tunnel plan, and the project begins. Resistance from the British public quickly ends the project, however.

1973   Construction begins on a new tunnel.

1975   The tunnel project is abandoned for economic reasons.

1984   British prime minister Margaret Thatcher and French president Francois Mitterrand announce they want to build a channel link.

1985   A design competition is held.

1986   The chunnel design is chosen.

1987   Work begins on the chunnel.

1990   The central service tunnel is completed.

1991   The northern tunnel and the southern tunnel are completed.

1993   The first test run of the chunnel is held.

1994   The chunnel opens to the public.

1996   Fire destroys part of the southern tunnel.

Glossary

debris—pieces of something that has been broken

engineer—someone who designs and builds structures, roads, or machines

estimate—a rough guess of how much something will cost

ferry—a ship that regularly carries people across a body of water

investment—money given to a company in the belief that more money will be earned in the future

laser—a narrow, powerful beam of light

maintenance—keeping something in good condition

petition—a letter signed by many people asking for something to be changed

proposal—a plan or idea

quarantine—to keep a person, animal, or plant away from others to stop the spread of disease

refugees—people who have been forced to leave home because of war, violence, or disaster

satellite—a spacecraft that orbits Earth and sends and receives signals

terminal—station

tunnel-boring machine—a huge drill that digs through rock to create a tunnel

ventilation—the system of sending fresh air into a place and taking stale air out

Books

Sandy Donovan, *The Channel Tunnel*. Minneapolis: Lerner, 2003.

Websites

The Channel Tunnel (www.raileurope.com). Includes a diagram of the tunnel and information about Eurostar.

Cool Stuff & the Incredible Feats of Construction: The Chunnel (www.constructmyfuture.com). This website includes the chunnel in its "Hall of Fame" and describes how it was built and how it is used today.

Great Engineering Feats: Channel Tunnel (www.teach ingtools.com). A brief overview of the chunnel's history and construction.

Welcome to Eurotunnel (www.eurotunnel.co.uk). The official Eurotunnel website, this site contains a photo gallery, interesting facts, and a history of the chunnel, plus train schedules, and ticket information.

Notes

1. Quoted in Drew Fetherston, *The Chunnel.* New York: Times Books, 1997, p. 96.
2. Quoted in Fetherston, *The Chunnel*, p. 19.
3. Quoted in Fetherston, *The Chunnel*, p. 295.

Index